YOUR KNOWLEDGE HAS VALUE

- We will publish your bachelor's and master's thesis, essays and papers

- Your own eBook and book - sold worldwide in all relevant shops

- Earn money with each sale

Upload your text at www.GRIN.com and publish for free

A. A. Ijagbuji, V. V. Schwarzkopf, I. I. Zakharov, D. B. Woods, T. C. Philips, K. M. Jackson, M. B. Saltzberg, B.V. Shevchenko, J. K. Johnson

Production of olefins via oxidative de-hydrogenation of C3 C4 fraction by CO2 over Cr Mo/MCM 41

GRIN Publishing

Bibliographic information published by the German National Library:

The German National Library lists this publication in the National Bibliography; detailed bibliographic data are available on the Internet at http://dnb.dnb.de .

Imprint:

Copyright © 2015 GRIN Verlag GmbH
Print and binding: Books on Demand GmbH, Norderstedt Germany
ISBN: 978-3-656-92242-1

This book at GRIN:

http://www.grin.com/en/e-book/294184/production-of-olefins-via-oxidative-de-hydrogenation-of-c3-c4-fraction

GRIN - Your knowledge has value

Since its foundation in 1998, GRIN has specialized in publishing academic texts by students, college teachers and other academics as e-book and printed book. The website www.grin.com is an ideal platform for presenting term papers, final papers, scientific essays, dissertations and specialist books.

Visit us on the internet:

http://www.grin.com/

http://www.facebook.com/grincom

http://www.twitter.com/grin_com

Production of olefins via oxidative de-hydrogenation of C_3-C_4 fraction by CO_2 over Cr-Mo/MCM–41

A.A. Ijagbuji[1*], V. V. Schwarzkopf[1, 2], I. I. Zakharov[1, 2], D. B. Woods[4], T. C. Philips[1], K. M. Jackson[4], M. B. Saltzberg[3], B. V. Shevchenko[3], J. K. Johnson[4]

[1]Institute of Technology, East Ukrainian National University, Severodonetsk, 93400, Ukraine.
[2]Boreskov Institute of Catalysis, Novosibirsk, Russia
[3] Moscow State University, Moscow, Russia
[4] University of Melbourne, Parkville, Victoria, Australia
e-mail address: dejiijagbuji@yahoo.com

Abstract

The present study investigates the oxidative de-hydrogenation of propane-butane (C_3-C_4) fraction over mono (Cr, Mo) and bi-metal (Cr-Mo) loaded MCM–41catalysts. The catalysts were prepared by sequential impregnation method at 500°C calcination temperature. Experiments were performed by feeding C_3-C_4 fraction and CO_2 into a continuous flow quartz reactor at atmospheric pressure (P = 1 atm.), reaction temperatures between 500 – 650°C, gas hourly space velocity within 100 – 400 h^{-1}, and at reaction time (t_r) = 2h. The physicochemical properties and performance of catalysts were evaluated by BET, XRD, H_2–TPR, and NH_3–TPD characterization techniques. The major products are *ethylene, propylene, isobutylene, butylene*. This paper reports that the total yield of olefins (Σ C_2-C_4) = 71.6 % was achieved at 89.5 % conversion level of C_3–C_4 at 630°C. The results indicate that the addition of Mo to Cr/MCM–41 modifies its catalytic activity. The Cr/MCM–41 and Mo/MCM–41 catalysts were prepared for comparison purposes.
Keywords: oxidative de-hydrogenation of C_3–C_4, C_3–C_4 conversion, selectivity to olefin, olefin yield, olefin production.

1. Introduction

Worldwide production of olefins exceeds that of any other chemicals and constitutes a sizable fraction of total petrochemical production. The chemical industry relies heavily on the low-cost and readily available saturated hydrocarbons as feedstock for many industrially significant processes. [1]
    Conventional technologies for olefins production involve catalytic de-hydrogenation of alkanes via either the steam cracking or fluid catalytic cracking method. However, these traditional methods for olefin production require high temperatures, high energy input, and are also limited by coke deposition and thermodynamic considerations.[2] Consequently, the existing commercial processes for olefins production *are very energy-intensive, exhaustive, and not cost-effective*. While these two routes are very well developed, increasing the capacity of these processes is only possible to some extent, as changing regulation limits the use of by-products (*notably aromatic molecules*) in fuels. For these reasons, the present industrial capacity for C_2-C_4 olefins production via these routes is expected to be insufficient, and therefore, cannot meet the fast-growing demand of olefins in the international market.[3]
    There are, however, a number of current challenges preventing oxygen assisted de-hydrogenation from being widely implemented. The difficulties inherent in oxidative dehydrogenation reactions revolve around selectivity control because all equivalent C-H bonds have

1

an equal bonding energy, and therefore an equal chance of reacting.[4] When two C-H bonds of neighboring carbons are split, a double bond is formed and alkanes are converted to alkenes. Thereby, oxygen addition to alkane feeds exposes the synthesized olefins to further oxidation conditions that results into the formation of *environmentally-damaging and economically-useless carbon oxides* (CO and CO_2), consequently decreasing the yield of alkenes [Eq. (3) – (4)]. Due to the exothermic nature of these reactions, it is expedient to remove heat and avoid the over-oxidation of alkanes to CO_2 in order to reach high selectivity to olefins.

$$C_3H_8 + {}^1\!/_2O_2 \rightarrow C_3H_6 + H_2O \qquad \Delta H_f^0 = -95.5 \ kJ/mol \qquad (1)$$
$$C_4H_{10} + {}^1\!/_2O_2 \rightarrow C_4H_8 + H_2O \qquad \Delta H_f^0 = -95.5 \ kJ/mol \qquad (2)$$
$$C_3H_6 + 3O_2 \rightarrow 3CO + 3H_2O \qquad \Delta H_f^0 = -219 \ kJ/mol \qquad (3a)$$
$$C_3H_6 + {}^9\!/_2O_2 \rightarrow 3CO_2 + 3H_2O \qquad \Delta H_f^0 = -1867.5 \ kJ/mol \qquad (3b)$$
$$C_4H_8 + 4O_2 \rightarrow 4CO + 4H_2O \qquad \Delta H_f^0 = -518 \ kJ/mol \qquad (4a)$$
$$C_4H_8 + 6O_2 \rightarrow 4CO_2 + 4H_2O \qquad \Delta H_f^0 = -2716 \ kJ/mol \qquad (4b)$$

Since producers seek to leverage their existing assets and the available internal streams to find an optimum solution for meeting the demands of olefins, oxidative de-hydrogenation of alkane with mild oxidant such as carbon dioxide, has been widely studied, as a potentially attractive route to circumvent the thermodynamic limitations, eliminate coking, and therefore, extend catalyst lifetime. Therefore, the design of effective catalytic systems that exhibit sufficient activity, high selectivity, periodically-regenerated under severe conditions, and yet operate at temperatures that minimize oxygenation of the desired products, are key performance demands for cost-effective production of olefins.

During the past decade, bi-metallic catalysts have attracted considerable attention because they show multiple functionalities, stability,[5 – 9] and selectivity[10 – 12] over mono-metallic catalysts. With respect to redox catalysis, chromium is one of the most important element to be incorporated, whereas, molybdenum exhibits excellent catalyst attrition resistance, facilitates easy products desorption from catalyst surface, maintain optical defect concentration, and blocks non-selective sites. A new class of mesoporous silica designated as MCM–41 has attracted increasing attention due to large surface area ($A_{BET} > 1000 \ m^2/g$), high pore volume ($V_p = 0.98 \ cm^3/g$), moderate hydrophobic character, as well as functionalized pore channels of large diameter ($d_p = 1.5$ to 10 nm) without network complications. The large surface area allows the binding of a large number of surface groups, whereas, the functionalized pore channels of large diameter allow an easy reaction with adsorbent. Pure siliceous hexagonal MCM–41 cannot be directly used as catalysts. It may be due to the limited thermal stability and negligible catalytic activity of this MCM–41, because of the neutral framework and the lack of sufficient acidity. Incorporation of hetero–atoms into a silica framework has been reported to increase the mechanical strength,[13] surface acidity of support, and enhance the overall catalytic performance.[14] Thereby, this new physicochemical properties derived from synergistic effects between bi-metals and the silica frame-work is highly desirable for catalytic applications.

To the best of our knowledge, in-depth investigation of MCM–41 supported mono Cr or Mo, and bi-metal Cr-Mo catalyst has not been previously reported in the literature, hence, this is deemed to be investigation worthy. In this paper, the oxidative de-hydrogenation of C_3-C_4 fraction into olefin, at temperatures (T) between 500 – 650°C, atmospheric pressure (P = 1 *atm.*), gas hourly space velocity (GHSV) = 100 – 400 h^{-1}, and reaction time (t_r) = 2h in a continuous quartz flow reactor is investigated using synthesized Cr, Mo, and Cr-Mo incorporated MCM–41 catalysts.

2. Experimental

2.1 *Materials and Methods*:

Tetra-methyl-ammonium hydroxide (TMAOH, 26 *wt.*%); acetic acid (CH₃COOH, 25 *wt.*%), distilled water (H_2O, $1dm^3$); chromium nitrate nonahydrate [$Cr(NO_3)_3 \cdot 9H_2O$, 99 %]; ammonium heptamolybdate ($[NH_4]_6Mo_7O_{24}.6H_2O$, 99 %); cetyltrimethylammonium-bromide (CTAB); and silica (SiO_2) were obtained from *Sigma-Aldrich*. The propane-butane fraction (C_3–C_4, 99.99 %) was supplied by *Naftogas*, Kiev; carbon dioxide (CO_2, 99.9 %) & nitrogen (N_2, 99.9 %) were obtained from *Azot chemical company*.

2.2 *Catalysts Preparation*:

The parent Si-MCM–41 was synthesized according to the following procedures: Typically, 3.0 g of SiO_2 was slowly added to a mixture of 2.63 g of aqueous TMAOH and 0.3 g of NaOH dissolved in 28.0 g of distilled water under vigorous stirring for 30 min. Subsequently, an aqueous suspension of CTAB (4.92 g in 14 g distilled water) was added to the above mixture, and the resultant mixture was stirred for 2 h. The gel mixture had a molar composition of SiO_2: 0.27 CTAB: 0.15 TMAOH: 0.15 NaOH: 60 H_2O. The pH of the resulting gel was adjusted to 10.3 by the drop-wise addition of acetic acid (CH₃COOH) followed by stirring for 3 h. The resulting gel was transferred into Teflon-lined stainless steel autoclaves (*ca.* 100 mL). The autoclave was placed in an oven and the synthesis gel was aged at 150°C for 24 h. After aging, the reactor was cooled down, the gel was filtered, washed extensively with distilled water in order to remove any unwanted species such as sodium ions, nitrate etc., and then dried at room temperature. Once dry, the gel was then calcined at 550°C (*heating rate of 1°C/min*) for 6 h to remove the organic template and create hollow porous structure. Prior to its application as a support, the calcined MCM–41 was characterized by low angle X-ray diffraction (XRD), N_2 adsorption, and transmission electron microscopy (TEM).

4 *wt.* % of Cr, V and Cr-Mo catalysts were prepared by incipient wetness impregnation of the powdered MCM–41 support:
(i) 0.3 g of chromium nitrate was dissolved in 30 mL de-ionized water, and then added to 1 g of MCM–41. After stirring for 0.5 h under reduced pressure via water aspiration at room temperature, the solid was filtered from the solution and dried at room temperature. Final calcination was carried out at 550°C for 6 h (*heating rate of 3°C/min*), and the Cr/MCM–41 catalyst was obtained; (ii) 0.092 g of ammonium heptamolybdate was dissolved in 30 mL de-ionized water, and added to 1 g of MCM–41. After stirring for 0.5 h under reduced pressure via water aspiration at room temperature, the solid was filtered from the solution and dried at room temperature. Final calcination was carried out 550°C for 6 h (*heating rate of 3°C/min*), and the Mo/MCM–41 catalyst was obtained; (iii) For 2 *wt.*% Cr and 2 *wt.*% Mo: 0.153 g of chromium nitrate and 0.046 g of ammonium heptamolybdate were dissolved in 30 mL de-ionized water, and added to 1 g of MCM–41. After impregnation, the resulting mixture was stirred for 0.5 h under reduced pressure at room temperature. The solid was filtered from the solution, dried at room temperature, and then calcined at 550°C for 6 h (*heating rate of 3°C/min*) to remove volatile impurities adsorbed on the catalyst surface. The other samples with different weight percentage loadings: 1.2Cr-2.8Mo/MCM–41 and 2.8Cr-1.2Mo/MCM–41 were prepared by similar way. Thereafter, the Cr/MCM–41, Mo/MCM–41, and Cr-Mo/MCM–41 catalyst samples were pressed and then sieved to appropriate sizes for catalytic evaluations. These catalysts were fully characterized by XRD, Infrared spectroscopy (IR), N_2 adsorption, and TEM.

2.3 Catalysts Treatment:

0.15 g of each catalyst was calcined in air at 550°C for 2 h to remove any volatile impurity adsorbed on the surface, followed by reduction in 10 % H_2/ 90% Ar at 500°C for 6 h to convert the catalyst into the metallic state. The total flow rate of the H_2/Ar mixture was $50cm^3$/min. The samples were then cooled at room temperature for 30 mins, and stored in inert atmosphere to avoid degradation.

2.4 Catalysts Test:

The catalytic activity of samples for the decomposition of C_3–C_4 fraction was investigated in a continuous flow quartz fixed-bed reactor (6 L vol.). The catalysts sample (0.15 g) was packed in the reactor and activated with flowing N_2 at 550°C for 2 h. After which the flow rate of reactant (C_3–C_4) and N_2 was maintained at 20 ml/min and 80 ml/min, respectively. The N_2 adsorption isotherms of calcined materials were measured at liquid nitrogen temperature (–196°C). At this temperature, because of its low cost and inert nature, nitrogen is an ideal adsorbent. The gas carrier was passed through a molecular sieve trap before being saturated with C_3–C_4. The gas product samples were analyzed by gas chromatograph.

2.5 Oxidative De-hydrogenation of C_3–C_4:

The C_3–C_4 feed composition was analyzed by gas chromatography «Chrom–5». It was established that the total composition equals 100 % by *volume*: propane = 20, *i*-butane = 60 and *n*-butane = 20. The oxidative dehydrogenation experiment of C_3–C_4 fraction was carried out in a continuous flow quartz fixed-bed reactor (5 L vol.). The experimental set up is shown in **Fig. 1**.

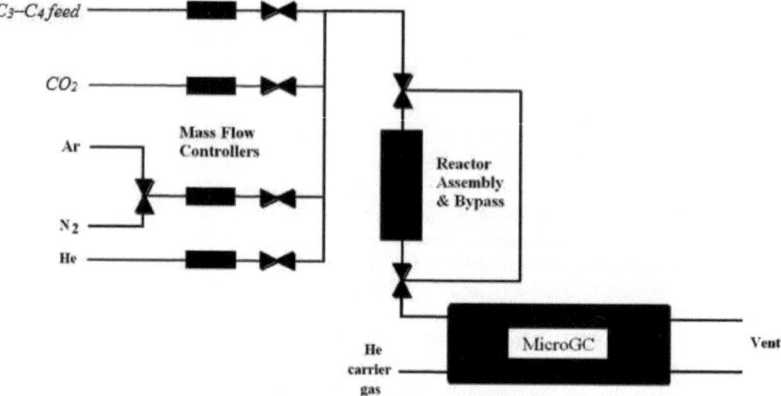

Fig. 1   *Schematic diagram of the pilot unit for dehydrogenation of C_3–C_4 fraction*

Before each run, the reactor was purged with N_2 for about 10 min and then de-coked using 15 % CO_2: 85 % N_2 mixture to ensure that the reactor walls and the coupon were coke free. This was accomplished by visually observing the appearance of the coupon through an observation hole in the furnace and by monitoring the weight of the coupon during the decoking process. If the appearance of the coupon was transparent and non-luminous, and its weight did not decrease with time, the coupon was assumed to be coke free. The reactor was again purged with N_2 for about 10 min, after which the hydrocarbon reactants and steam were introduced. The primary reason for N_2 purge before and after decoking experiments was to minimize the accumulation of potentially explosive mixtures in the reactor. Each run was repeated at least five times to ensure reproducibility and to assess the range of experimental errors associated with the experiments.

In order to determine the catalytic specie with the best performance, three series of experiment were performed. In the 1st experiment, C_3-C_4–CO_2–N_2 mixture on the Cr/MCM–41 was fed into the reactor. The propane-butane fraction, carbon dioxide, and nitrogen were supplied from a pressure cylinder through reduction valves. The reactant gases consisting of C_3–C_4, CO_2, and some additional N_2 carrier gas were then mixed with steam and transported to the reactor through electrically heated lines at a flow-rate desired ($\pm$ 5%) for the given experiment. A molecular sieve was used at the entrance of the reactor with the objective of retaining impurities coming from the feeding line. The flow rate of C_3–C_4 fraction, CO_2, and N_2 was regulated by mass flow controller that was calibrated before the experiments. The reactant mixture was subjected to thermal treatment at temperature range within $T = 500 - 650°C$, atmospheric pressure ($P = 1$ atm.), and for (t_r) = 2 h total reaction time. The flow rate of reactant mixture (C_3–C_4/CO_2/N_2 = 20: 46: 34) through the reactor was 100 cm^3/min, and 1 L/min for water. The average residence time of reactant mixture in the reactor was about 10 s. The gas hourly space velocity of $100 - 400$ h^{-1} was varied to obtain different level of C_3–C_4 conversion. The catalyst activity was maintained by regeneration after every experimental hour using a N_2/CO_2 mixture. During the regeneration, the output stream from the reactor was checked for carbon dioxide. The reaction products were obtained at the bottom of reactor, cooled, separated into individual components, and analyzed at 10 min. intervals by gas chromatograph–mass spectrometer (GC–MS).In the 2nd experiment, C_3-C_4–CO_2–N_2 mixture on the Mo/MCM–41 was fed into the reactor. In the 3rd experiment, C_3-C_4–CO_2–N_2 mixture on Cr–Mo/MCM–41 was fed into the reactor. The same experimental condition described in 1st experiment was applied for the 2nd and 3rd experiments.

3. Results and Discussion

3.1 Catalysts Characterization:

The Textural characteristics such as metallic composition, BET surface area and pore distributions of catalyst samples are compiled in **Table 1**. Atomic absorption spectroscopy by «Accu-sorb» was used to measure the elemental composition of catalysts. The BET surface areas (determined by N_2 physi-sorption) and the pore size distributions of all the catalyst samples were calculated using the *Brunauer-Emmett-Teller* (BET) and the *Barrett-Joyner-Halenda* (BJH) methods, respectively. The N_2 adsorption-desorption isotherms on catalyst samples show a characteristic capillary condensation pore-filling step of a typical reversible type IV adsorption isotherm as defined by IUPAC for mesoporous materials.[15] After calcinations, the observed surface area of parent MCM–41 (1238.6 m^2/g) may be attributed to the removal of organic template, and CTABr from the material which, consequently, resulted in an increase in the adsorption site for the nitrogen molecules. Upon Cr, Mo, and Cr-Mo incorporation on MCM–41 support, it can be seen that the surface areas of catalyst samples show a similar trend: being maximum for the 1.2Cr-2.8Mo/MCM–41 impregnated sample

(1098.4 m^2/g), and minimum for the 4Mo/MCM–41 (932 m^2/g) upon calcination. The pore size distribution curves of samples as calculated by BJH method centered between $V_p = 0.652 - 0.948$ cm^3/g (pore volume) and $d = 2.6 - 3.2$ nm (pore diameter) are presented in **Figure 2**. Nevertheless, the observed slight reductions in textural characterization, in comparison to values of the parent MCM–41, may be attributed to modification of the pore wall with metal precursor which reduces the scattering contrast between the pores and the walls of the molecular sieve.

Table 1 *Physicochemical properties of the different catalyst samples*

Catalysts	Composition (wt.%)		Surface area (m^2/g)	Pore volume, V_p (cm^3/g)	Pore diameter, d (nm)
	Cr	Mo			
Si-MCM–41	–	–	1238.6	0.948	3.2
4Cr/MCM–41	4	–	949	0.673	2.7
4Mo/MCM–41	–	4	932	0.652	2.6
1.2Cr-2.8Mo/MCM–41	1	3	1098.4	0.780	2.9
2Cr-2Mo/MCM–41	2	2	975	0.695	2.8
2.8Cr-1.2Mo/MCM–41	3	1	966	0.719	2.9

Further, the differences shown between the mono and bi-metallic catalysts are often related to the different conditions that metal precursors undergo during the impregnation process i.e. on the one hand, the $Cr(NO_3)_3 \cdot 9H_2O$ behaves as a strong acid (with a pH around 2.5), whereas, the $([NH_4]_6Mo_7O_{24}.6H_2O)$ shows a weak acid character (with a pH around 5.6); on the other hand, the aqueous mixture of $Cr(NO_3)_3 \cdot 9H_2O$ and $([NH_4]_6Mo_7O_{24}.6H_2O)$ forms a Cr-Mo/MCM–41 bi-metallic catalyst with an average pH of 4.3. It has been established that at pH values between 4.0 to 5.0, a hydrolytic reaction occurs in $Cr(NO_3)_3 \cdot 9H_2O$ to form colloidal $Cr(OH)_2$ species, whereas, $(Mo_7O_{24}^{6-})$ continue being the dominant species during impregnation process. The simultaneous interaction of Cr and Mo species with a positive electrically-charged surface of silica propitiates the selective adsorption of $(Mo7O_{24}^{6-})$ anions on the surface, followed by precipitation of Cr particles during evaporation of the precursor. Consequently, the negative influence of diffusive retardation and re-adsorption inside pores are suppressed to a maximum extent. It should be re-called that calcination of Si-MCM-41 containing large amounts of carbonaceous species can leave carbon deposits or coke as contaminant in the pores and pore blocking may occur. The high values for the surface area and pore volumes of catalysts from the data presented in **Table 1** indicate that no pore blocking has occurred after catalyst preparation with both precursors. In addition to that, neither the shrinkage of the average pore diameter, nor a broadening of pore size distribution was observed.

Sintering can be understood using the *Gibbs–Thompson* relationship where larger particles with lower chemical potential will grow at the expense of smaller particles with higher chemical potential: the driving force being the reduction of the total surface energy of the system. With these considerations, metals with an electro-negativity $\chi < 1.5$ on the Pauling scale should react with a SiO$_2$ substrate. However, sintering was not observed in both mono-functional catalysts, owing to the relatively high electro-negativity values of χ Cr = 1.66 and χ Mo = 2.16. Further, sintering of the tubular structure of MCM–41 support, as well as, the transition metal active sites was not observed. Though slight reduction of specific surface area and pore volume was noticed for the catalyst after reaction but values are in tandem with those typical for mesoporous materials. [15]

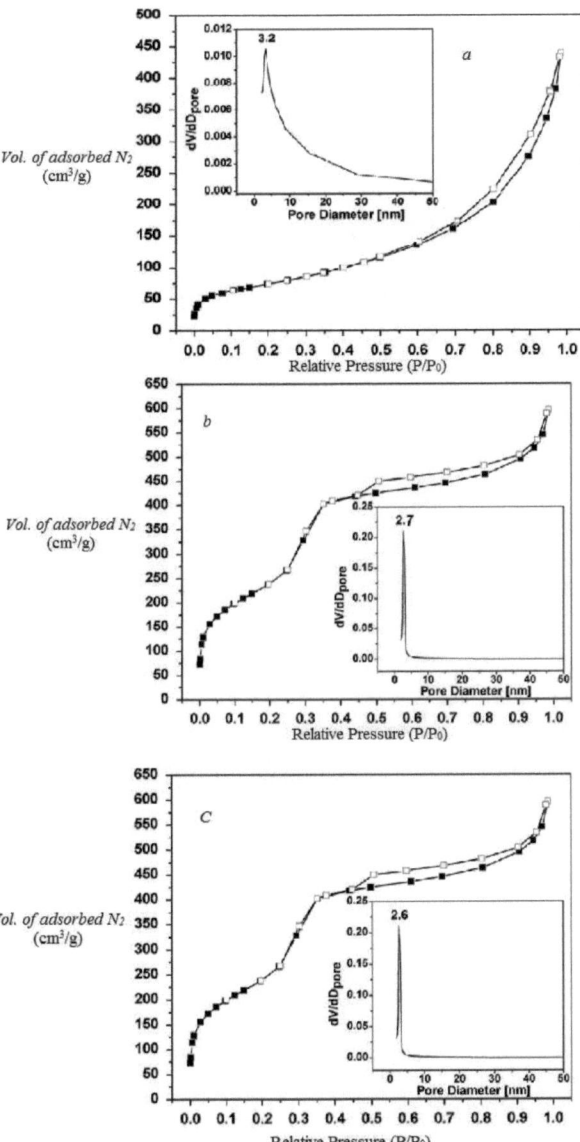

7

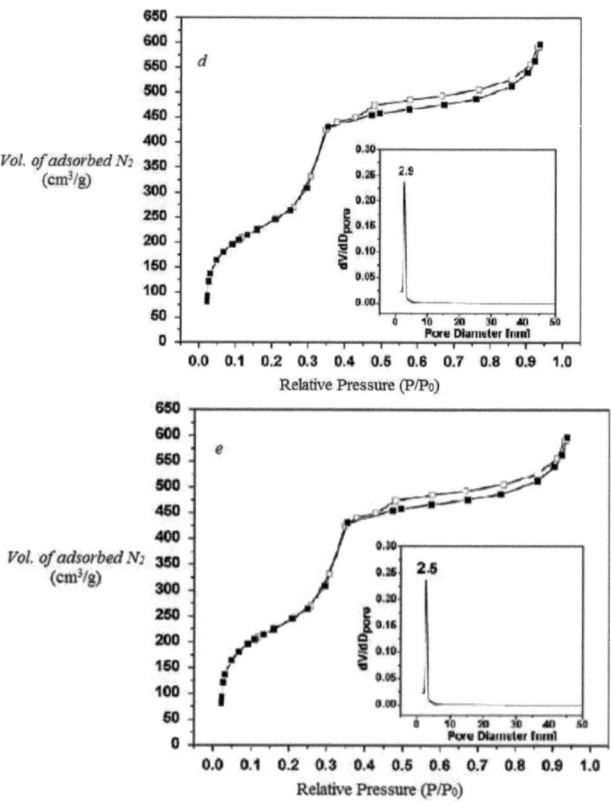

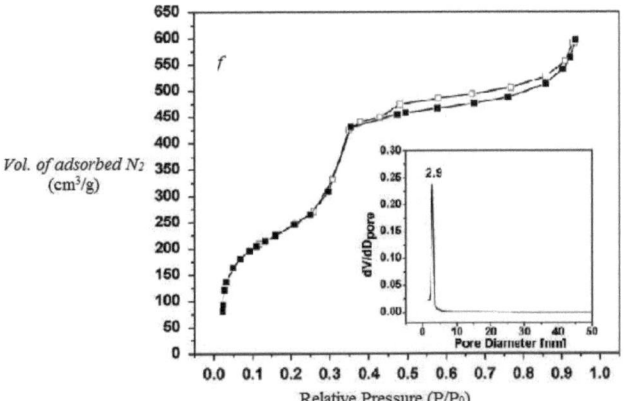

Fig. 2 *N₂ adsorption-desorptiion isotherm of catalysts* (a) Si-MCM–41, (b) 4Cr/MCM–41, (c) 4Mo/MCM–41, (d) 1.2Cr-2.8Mo/MCM–41, (e) 2Cr-2Mo/MCM–41, (f) 2.8Cr-1.2Mo/MCM–41.

3.2 x-ray diffraction of calcined catalysts:

The X-ray diffraction patterns obtained for siliceous MCM–41 and the impregnated samples after calcination in air at 550°C was studied using X-ray diffractometer (XRD), and the results are shown in **Fig. 3**. The X-ray tube was operated at 80 kV and scanning rate of 10°/min. The K_α radiation of diffracted beam mono-chromator was selected and angular range from 10° to 60° was recorded using step scanning. After the removal of surfactant template via the calcination stage, the low angle reflection intensities increase and the mesoporous ordered structure could be dramatically affected. In the low 2-theta region of 1° – 10°, four distinct diffraction peaks are visible which can be indexed as (100), (110), (200) and (210) reflections, respectively. This is characteristic of hexagonal mesoporous MCM–41 structure.[16] It can be seen that an MCM–41 support material with excellent ordering was prepared. The parent MCM–41 exhibits a narrow and strong peak at 41.03 Å which corresponds to a lattice parameter, a_o = 47.03 Å and other three weak peaks at 23.76 Å, 20.55 Å and 15.74 Å due to the (100), (110), (200) and (210) reflections, respectively. This may be attributed to the removal of organic surfactant template and condensation of silanol groups in the pore walls. However, it is worthy to mention that organic template removal may lead to the rearrangement of silica walls of siliceous MCM–41, thereby, increasing the crystallinity of MCM–41. Comparatively, the mono-metal 4Cr/MCM–41 and 4Mo/MCM–41 impregnated catalyst samples show well resolved four diffraction lines at 2θ range = 2.4, 4.3, 4.8 and 6.6 indexed to (100), (110), (200) and (210) peaks corresponding to hexagonal symmetry (**Fig. 3**(b and c)). In the case of bi-metal impregnated samples, three diffraction lines are observed indicating well ordered mesoporous support (**Fig. 3**(d–f)). The absence of crystalline α-Cr_2O_3 and Mo_2O_5 at a range between 2θ = 10 – 60° for all the samples show the highly disbursed metal oxide phase over the support. In addition, the structural integrity of the support material is retained upon the bi-metallic catalyst impregnation with $Cr(NO_3)_3·9H_2O$ and ($[NH_4]_6Mo_7O_{24}.6H_2O$) complex precursors. Moreover, the 4Cr/MCM–41, 4Mo/MCM–41, 2.8Cr-1.2Mo/MCM–41, 2Cr-2Mo/MCM–41, and 1.2Cr-2.8Mo/MCM–41 impregnated catalyst samples show high orderliness regardless of the metal loading.

9

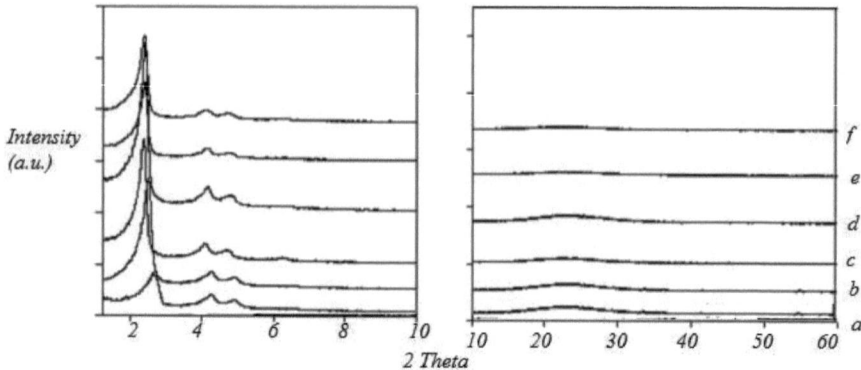

Fig. 3 *XRD data of the calcined catalysts* (*a*) Si-MCM–41, (*b*) 4Cr/MCM–41, (*c*) 4Mo/MCM–41, (*d*) 2.8Cr-1.2Mo/MCM–41, (*e*) 2Cr-2Mo/MCM–41, (*f*) 1.2Cr-2.8Mo/MCM–41.

3.3 Temperature-programmed analysis of calcined catalysts:

The oxidative dehydrogenation reaction of alkanes over transition-metal oxides supported catalysts, proceeds via redox mechanism, in which the reducibility is a key factor for the catalyst activity. [17]

Characterization of MCM–41 with x-ray diffraction yields a diffractogram with a limited number of reflections, all situated at low angles, hence, x-ray diffraction is known to be relatively insensitive for the detection of small crystalline nanoparticles.[18] Therefore, the materials were characterized by H_2–TPR. **Fig.** 3(a–f) shows the reducibility of Cr, Mo or Cr-Mo impregnated MCM–41 support. The TPR results are presented in **Table 2**.

Table 2 The temperature programmed analysis (H_2–TPR and NH_3–TPD) of Cr, Mo and Cr-Mo supported MCM–41 catalysts.

Catalysts	TPR		TPD
	$T_M(°C)$	H_2 uptake (mmol/g)	Total acidity (mmNH$_3$/g)
Si-MCM–41	–	–	0.080
4Cr/MCM–41	368, 444	0.246	0.339
4Mo/MCM–41	539, 623	0.639	0.321
1.2Cr-2.8Mo/MCM–41	434, 466, 498	0.788	0.540
2Cr-2Mo/MCM–41	377, 460, 492	0.612	0.433
2.8Cr-1.2Mo/MCM–41	373, 451	0.508	0.544

Fig. 4 shows H_2–TPR and NH_3–TPD profiles of MCM–41 supported mono Cr, Mo and Cr-Mo bi-metallic catalysts. It can be seen that no reduction peak was recorded for the Si-MCM–41. For the 4Cr/MCM–41 catalyst, the TPR profile has two hydrogen consumption peaks at 368 and 444°C, respectively. Ellison et al.[19] attributed such peaks to CrO_3 clusters with smaller interaction over the silica support. However, the H_2 consumption was found to be smaller than for other samples.

For the 4Mo/MCM–41 catalyst, two hydrogen consumption peaks at 539 and 623°C, respectively. The reduction peak maxima between 500 and 550°C is attributed to the reduction of surface Mo^{5+} species to surface Mo^{3+} species.[20–22] The presence of a less intense peak between 580 and 630°C for 4Mo/MCM–41 is attributed to the reduction of polymeric and bulk molybdena species.[21] Because molybdenum silicates are difficult to be reduced at reaction temperature, Mo-based catalysts have less active species, and hence present lower C_3–C_4 de-hydrogenation activities.

In the case of bimetal loading, for 1.2Cr-2.8Mo/MCM–41 sample, a shift in the onset of reduction peak towards lower temperature between 340 and 523°C compared to 4Mo/MCM–41 indicating more reducible species due to the Cr-Mo oxide interaction leading to mixed oxide phase over the support. A small reduction peaks at 434°C and a distinct reduction band at 466°C were observed. The presence of high temperature peak at 498°C, which extends up to 540°C, can be due to monomeric or low oligomeric surface dispersed tetrahedral species. In addition, it can be pointed out that the MoO_x species in the presence of CrO_x species tends to inhibit the agglomerization of molybdena species, and hence, the formation of bulk-like molybdena species becomes less prominent. The peaks at 373 and 451°C became dominant for the samples 2Cr-2Mo/MCM–41 and 2.8Cr-1.2Mo/MCM–41 with increasing chromium content indicating poly-chromates formation. Airaksinen et al.[23] observed that such poly-chromates species on chromia/alumina sample are more reducible than mono-chromates species.

The H_2 consumptions (mmol/g) are in the sequence: 1.2Cr-2.8Mo/MCM–41 (0.788) > 4Mo/MCM–41 (0.639) > 2Cr-2Mo/MCM–41 (0.612) > 2.8Cr-1.2Mo/MCM–41 (0.508) >> 4Cr/MCM–41 (0.246) > Si-MCM–41 (0), respectively. The depreciation of H_2 consumption with high Cr loading indicates increasing Cr-Mo oxide interactions, which may lead to the formation of mixed oxide phases and, therefore, may inhibit the thermal decomposition of chromium species in oxidation states VI and/or V during calcination.[24–25] The amount of transition metal reduced for all the catalysts were calculated. In the case of mono-metal impregnated Si-MCM–41 such as 4Mo/MCM–41 and 4Cr/MCM–41, a reduction of 46 and 88.3 % of metals are observed indicating reduction of Mo^{5+} to Mo^{3+} and Cr^{6+} to Cr^{3+} species, respectively. In the case of bi-metal impregnated MCM–41, reduction temperature was observed in between Mo or Cr impregnated MCM–41 samples. Almost all the metal content in 1.2Cr-2.8Mo/MCM–41 is observed to be reduced to trivalent state, whereas the reduction of 79 and 63.6 % occurs over 2Cr-2Mo/MCM–41 and 2.8Cr-1.2Mo/MCM–41, respectively.

It has been reported that the acid-base functionality of catalyst plays a significant role in the selectivity of dehydrogenation products.[26–27] The NH_3 consumptions (mmNH₃/g) follow the sequence: 1.2Cr-2.8Mo/MCM–41 ~ 2.8Cr-1.2Mo/ MCM–41 > 2Cr-2Mo/MCM–41 > 4Cr/MCM–41 > 4Mo/MCM–41 > SiMCM–41. The increased surface acidity shows the stronger interaction of mixed oxides over support than that of the mono metal oxide supported catalysts. Though TPD cannot provide direct information about interaction energies of ammonia molecule with catalyst's surface, the increased amount of desorbed NH_3, in our view-point, may be due to the existence of synergistic effect between the metal oxides i.e. increase in the number of adsorption sites, or a change in stoichiometry of the adsorption adduct.

3.4 Oxidative de-hydrogenation of C_3–C_4:

During oxidation, the alkane molecule is adsorbed to the surface of oxygen atom [Eq. (5)], followed by the cleavage of C–H bond to form an alkyl intermediate, and a hydroxyl group on the catalyst surface [Eq. (6)]. The adsorbed alkyl specie looses a second hydrogen atom, thus, forming alkene, and another hydroxyl group on the catalyst surface [Eq. (7)]. Two hydroxyl groups combine to form water and lattice vacancy [Eq. (8)]. The oxidative de-hydrogenation reaction is postulated to occur via a redox cycle, where the catalysts lattice oxygen takes part in the oxidation reaction [Eq. (9)], and then the reduced catalyst is re-oxidized following a *Mars-van Krevelen* mechanism.[28]

$$C_nH_{2n+2} + O^* \leftrightarrow C_nH_{2n+2}O^* \quad (n = 3, 4) \tag{5}$$
$$C_3H_8 + O^* \leftrightarrow C_3H_8O^* \tag{5a}$$
$$C_4H_{10} + O^* \leftrightarrow C_4H_{10}O^* \tag{5b}$$
$$C_nH_{2n+2}O^* + O^* \rightarrow C_nH_{2n+1}O^* + OH^* \quad (n = 3, 4) \tag{6}$$
$$C_3H_8O^* + O^* \rightarrow C_3H_7O^* + OH^* \tag{6a}$$
$$C_4H_{10}O^* + O^* \rightarrow C_4H_9O^* + OH^* \tag{6b}$$
$$C_nH_{2n+1}O^* \rightarrow C_nH_{2n} + OH^* \quad (n = 3, 4) \tag{7}$$
$$C_3H_7O^* \rightarrow C_3H_6 + OH^* \tag{7a}$$
$$C_4H_9O^* \rightarrow C_4H_8 + OH^* \tag{7b}$$
$$2OH^* \leftrightarrow H_2O + V^* + O^* \tag{8}$$
$$O_2 + 2V^* \rightarrow 2O^* \tag{9}$$
$$C_3H_8 + CO_2 \leftrightarrow C_3H_6 + CO + H_2O \tag{10}$$
$$C_4H_{10} + CO_2 \leftrightarrow C_4H_8 + CO + H_2O \tag{11}$$
$$2C_3H_6 \leftrightarrow C_4H_8 + C_2H_4 \tag{12}$$
$$4C_3H_8 + 4CO_2 \leftrightarrow 3C_4H_8 + 4CO + 4H_2O \tag{13}$$
$$C_3H_8 + 3CO_2 \leftrightarrow C_2H_4 + 4CO + 2H_2O \tag{14}$$
$$C_3H_8 \leftrightarrow C_3H_6 + H_2 \tag{15}$$
$$C_2H_4 + CO_2 \leftrightarrow CH_4 + 2CO \tag{16}$$

The catalytic performances of samples, as well as the effect of external mass transfer on C_3–C_4 oxidation was studied by varying stirrer speed from $100 - 400$ h^{-1}, T $= 500 - 650°$C, and P $= 1$ atm. Since the pore structure of catalyst surface vary at different reaction temperatures, any impurity generated at different temperatures blocks the catalyst pore structure, resulting in lower catalytic activity and shorter catalyst effective time. Therefore, five different temperatures ranging from 500 to 650°C were selected to investigate the optimum reaction temperature. The optimum reaction temperature was therefore, carefully determined by CO_x removal efficiency and effective time of catalyst. The best experimental result was achieved at a gas hourly space velocity of 250 h^{-1}, which is presented in **Table 3**. *Gas hourly space velocity is the gas speed during a reaction and affects various catalytic reactions differently* i.e. a low *gas hourly space velocity* is more suitable for catalysts with a dense structure, whereas high *gas hourly space velocity* is effective for catalysts with a loose structure. In this study, five values of *gas hourly space velocity* at 100 h^{-1}, 200 h^{-1}, 250 h^{-1}, 300 h^{-1}, and 400 h^{-1} were used to evaluate the effects on C_3–C_4 *conversion.*

Table 3 *Typical catalytic performances in propane-butane oxidative conversion*

Catalysts	T (°C)	Conv. (%) C3-C4	Selectivity (%)					Yield (%)				
			C2H4	C3H6	iC4H8	CH4	C2H6	C2H4	C3H6	iC4H8	CH4	C2H6
4Cr/M–41	500	44.0	23.0	18.8	15.4	–	–	10.1	8.3	6.8	–	–
	550	52.0	27.7	17.6	13.5	–	–	14.4	9.2	7.0	–	–
	600	65.9	30.0	14.5	9.9	–	–	19.8	9.6	6.5	–	–
	630	70.0	32.5	10.8	7.8	–	–	22.8	7.6	5.5	–	–
	650	75.2	33.8	9.4	4.6	27.5	13.8	25.4	7.1	3.5	20.6	10.4
4Mo/M–41	500	43.3	24.8	19.6	16.9	–	–	10.7	8.5	7.3	–	–
	550	53.5	28.0	16.2	12.9	–	–	15.0	8.7	6.9	–	–
	600	68.1	30.7	15.5	10.1	–	–	20.9	10.6	6.9	–	–
	630	73.8	35.6	12.9	8.4	–	–	26.3	9.5	6.2	–	–
	650	77.5	39.4	8.2	6.3	29.8	14.2	30.5	6.4	4.9	23.1	11.0
1.2Cr-2.8Mo/M–41	500	61.0	28.9	35.0	14.2	–	–	17.6	21.4	8.7	–	–
	550	69.3	37.4	32.8	12.3	–	–	26.0	22.8	8.5	–	–
	600	78.8	39.5	30.6	10.9	–	–	31.1	24.1	8.6	–	–
	630	89.5	42.8	28.9	8.3	–	–	38.3	25.9	7.4	–	–
	650	93.4	43.3	20.0	5.0	21.1	10.6	40.4	18.7	4.7	19.7	9.9

P = 1 atm., GHSV = 250 h^{-1}, C3–C4 : CO2 = 1 : 1.5, and reaction time (t$_r$) = 2h.

The feed conversion (X), olefin selectivity (S) and olefin yield (Y) were calculated using following relationship:

Conversion (%) = (content of C3-C4 feed) – (content of C3-C4 in product)
                                    content of C3-C4 feed

Selectivity of olefins (%) = (% content of olefins in product) x 100
                                    C3–C4 conversion (%)

Yield of olefins (%) = (% C3–C4 conversion) x (% olefin selectivity)

Table 3 shows that when the *gas hourly space velocity* = 250 h^{-1}, the C3–C4 oxidation reached a conversion level of 93.4 % and the effective time was 10 min. With increased *gas hourly space velocity* from 250 h^{-1} to 400 h^{-1}, the *oxidative conversion* rate of the catalyst significantly decreased. The removal rate also decreased when *gas hourly space velocity* = 100 h^{-1}. Thus, the optimum GHSV was determined to be 250 h^{-1}. When the *gas hourly space velocity* was increased, the C3–C4 oxidation efficiency increases due to decrease of the out-diffusion. When the *gas hourly space velocity* was further increased, the *conversion level* increases correspondingly, suggesting that an increase of *gas hourly space velocity* resulted in increase in gas conversion but shortened retention time of gas, thus, reaction tends to be incomplete. From this point of view, the proper *gas hourly space velocity* was of great importance for *propane-butane oxidative de-hydrogenation.*

From the data presented in **Table 3**: (i) When the reaction temperature was varied from 500 to

650°C on the 4 *wt.*% Cr/M–41 catalyst, it can be seen that the selectivity to olefin slightly decreases (from 57.2 to 47.8 %) with increase in C_3-C_4 conversion level (from 44.0 to 7.2 %). A rise in the total yield of olefins (Σ C_2-C_4) was observed from 25.2 to 36.0 %. (ii) When the reaction temperature was varied from 500 to 650°C on the 4*wt.*% Mo/Cr/M–41 catalyst, the selectivity to olefin slightly decreases from 61.3 to 53.9 %. Nevertheless, C_3-C_4 conversion level increases from 43.3 to 77.5 %, as does the total yield of olefins (Σ C_2-C_4) from 26.5 to 41.8 %; (iii) When the reaction temperature was varied from 500 to 650°C on the 1.2Cr-2.8Mo/M–41 catalyst, the C_3-C_4 conversion level increases from 61.0 to 93.4 %, as does the total olefins yield (Σ C_2-C_4) from 47.7 to 71.6 % (at 630°C), but followed by a sharp reduction to 63.8% at 650°C.

It can be seen from this result that the C_3-C_4 conversion level and selectivity to ethylene formation increases with an increasing temperature. On the contrary, selectivity to propylene and butylenes decreases with increase in both reaction temperature and C_3-C_4 conversion level. The corresponding decrease in C_3-C_4 alkenes selectivity at high C_3-C_4 conversion level may be attributed to higher reactivity of C_3 and C_4 olefins than the reactant mixture (C_3-C_4 with CO_2) over the catalysts. It has been reported that activation energy requirements for dehydrogenation decreases with the increase in carbon chain.[29] For instance, butane is expected to be more quickly cracked over bi-metallic silica support comparative to propane i.e. while butane is partially cracked to either propylene and methane ($C_4H_{10} \rightarrow C_3H_6 + CH_4$) or to ethylene and ethane ($C_4H_{10} \rightarrow C_2H_4 + C_2H_6$); propane, is partially cracked to ethylene and methane ($C_3H_8 \rightarrow C_2H_4 + CH_4$). In addition to that, formation of ethylene may occur due to secondary cracking of butene isomers via primary carbenium ion formation. Nonetheless, propylene molecule (C_3H_6) has the tendency to form allylic site (delocalized unpaired electron) or weak C-H links that are liable to secondary reaction, consequently, decreasing its yield. As a result, it is not feasible to get higher yield or selectivity to propylene/butylene at increased temperature. Nevertheless, at higher conversion level of C_3–C_4, the particularly low selectivity to C_4 alkenes is due to hydride transfer reaction activation, and also suggests C_4 alkanes to be the dominant reaction in the investigated system. Moreover, it was observed that increase in the reaction temperature beyond 630°C has undesirable effect on olefin selectivity i.e. the formation of cracking products such as C_1, and C_2-hydrocarbons were observed at more pronounced quantities from the reaction products. The content of methane increasing with increasing the temperature indicates that the splitting reactions of carbon compounds, accompanied by *in situ* formation of hydrogen, and its subsequent reaction with carbon dioxide, are mostly preferred. The results obtained allow us to speculate about a potential suitability of carbon dioxide, which can serve not only as the reactant, but also as a heat supporter and together with water vapor, as a dilutant of the feed for a mild-temperature pyrolysis.

It is interesting to observe that the catalysts modified with bi-metals over MCM–41 shows an enhanced de-hydrogenation activity. This improved results of bi-metal impregnated MCM–41 samples can be explained to the synergetic effect of active metals (CrO_x and/or MoO_x species), which according to the results of XRD and TPR are well dispersed on the surface of mesoporous support. The selectivity to de-hydrogenation products over Cr–Mo–O based MCM–41 can be related to the acid-base character of the catalysts. It may be suggested that the presence of MoO_x species favor dehydrogenation while Mo over layer or Mo_2O_5 crystallites play detrimental role by favoring coke formations. The observed high conversion of 1.2-Cr2.8Mo/MCM–41, 2.7Cr-1.2Mo/MCM–41, and 2Cr-2Mo/MCM–41 shows the presence of dispersed CrO_x and MoO_x species that play an important role for the formation of key intermediate species. Hence, the overall yield of olefins over bi-metal catalyst, apparently, gets an enhancement in comparison to values obtained for mono-metallic catalysts. The enhancement of surface acidity of silanol groups and the higher facile reducibility as a result of strong interaction between Cr-Mo and MCM–41 support may be the reason for improved results of C_3–C_4 conversion and olefins yield with the bi-metallic supported catalysts. Therefore, from the XRD result in the oxidative dehydrogenation of C_3–C_4 over transition metal oxide supported catalyst: our perspective on reducibility as a key factor for catalyst activity agrees with study.[30] The catalytic evaluation shows that 1.2-Cr2.8Mo/MCM–41 shows maximal result in both olefins selectivity (80 %), and total yield of olefins (Σ C_2-C_4) = 71.6 % at C_3–C_4 conversion level (89.5 %) under reaction temperature T = 630°C, P = 1 atm., and gas hourly space velocity = 250 h^{-1}. Under the mild treatment, the catalyst was tested for two consecutive runs, and the catalyst activity was found to be stable. Thereby, our perspective on the tremendous catalytic influence of MCM–41 supported bi-functional catalyst in oxidative de-hydrogenation reaction is broadly supportive of the previous study.[13 – 14] Moreover, the Cr-Mo/MCM–41 shows higher catalytic activity, selectivity, as well as yield compared to the values previously reported in work.[31]

At reaction temperatures between T = 500 – 630°C, in-situ spectroscopic analysis of the reaction products was performed using gas Chromatograph «Chrom–5». The major oxidation products are *ethylene, propylene, isobutylene*, and *water.* In addition to that, neither cracking products (such as C_1, C_2-hydrocarbons), aromatic compounds (e.g. C_6H_6), nor oxygenated compounds (e.g. R–CHO or CH_3OH) were present in the exhaust of the reactor. Since C_3 and C_4 olefins are destroyed through subsequent reactions (secondary cracking), it is likely that an optimized plant design, with strong integration between reaction and cooling zones, will reduce the incidence of this undesired effect. Though, the partial contribution to olefins via heterogeneous de-hydrogenation or oxidative dehydrogenation reactions cannot be excluded, the strong dependence of all hydrocarbon products on feedstock conversion (directly related to the operating temperature) is a clear indication that homogeneous reactions plays a major role.[32]

Nevertheless, it is imperative to clearly state that the maximum overall yield of olefins is dependent on a number of parameters such as catalysts pre-treatment/preparation, the choice of metal precursor, the type of support, the nature and amount of oxidant used, as well as the operating conditions (e.g. reaction temperature, regeneration-dehydrogenation cycles, reactant flow-rate, etc.).

4. Conclusion

The oxidative dehydrogenation of propane-butane fraction to olefins in the presence of oxygen over mono (Cr/MCM–41 or Mo/MCM–41) and bi-metal (Cr-Mo)/MCM–41 catalysts, at temperatures T = 500 – 650°C, atmospheric pressure (P = 1 atm.), and gas hourly space velocity = 100 – 400 h^{-1}, has been investigated in a continuous flow quartz reactor. From this study, it can be concluded that:
(i) The oxidative dehydrogenation at a molar ratio of C_3–C_4/CO_2 = 1: 1.5, T = 630°C, P = 1 atm., and gas hourly space velocity (GHSV) = 250 h^{-1} over (Cr-Mo)/MCM–41 catalyst are optimal conditions for obtaining the highest selectivity (80 %) and yield of olefins (ΣC_2-C_4 = 71.6 %);
(ii) The 1.2%wt.Cr–2.8%wt.Mo/MCM–41 catalyst showed a better tolerance to carbonaceous deposits compared to the mono-metallic catalysts (4 %wt. Cr/MCM–41 and 4 %wt. Cr/MCM–41);
(iii) The presence of Mo enhances the catalytic performance of Cr-Mo/MCM–41 for the selective oxidative de-hydrogenation of C_3–C_4;
(iv) The catalytic efficiency on C_3–C_4 conversion, selectivity to olefins, and olefins yield are in the sequence; 1.2%wt.Cr–2.8%wt.Mo/MCM–41 >> 4 %wt. Mo/MCM–41 > 4 %wt. Cr/MCM–41;
(v) The reactants (e.g. C_3–C_4, and CO_2) are inexpensive and abundantly available, thus, making the overall production process to be cost-effective.

Conclusively, the oxidative de-hydrogenation of propane-butane fraction in the presence of carbon dioxide is environmentally benign technology and can, therefore, be considered as a useful guidance to design an industrial plan for the conversion of low molecular weight paraffin hydrocarbons to olefins; however, this novel technology will be ready for the industrial application only after completing an extensive experimental work in the range of operating conditions identified as the most promising by technical-economic evaluations, and after improving the catalyst and reactor design reliability with respect to heat management.

Acknowledgements

This work was supported by the *State Foundation for Basic Research*. The co-operations of *Prof.* B.C. Scott and *Prof.* C. W. Anderson are gratefully acknowledged.

References

[1] B.M. Weckhuysen, I.E. Wachs, R.A. Schoonheydt, *Chem. Rev.* 96 (1996) 3327.

[2] Chan, K.Y.G., F. Inal, and S. Senkan, Suppression of coke formation in the steam cracking of alkanes: ethane and propane, *Industrial & Engineering Chemistry Research*, 1998 37 (3) 901 – 907.

[3] Corma, A.; Melo, F.V.; Sauvanaud, L.; Ortega, F. *Catal. Today,* 699 (2005) 107 – 108.

[4] J.E. Germain, *Catalytic Conversion of Hydrocarbons*, Academic Press, London, 1969.

[5] V.R. Stamenkovic, B. Fowler, B.S. Mun, G. Wang, P.N. Ross, C.A. Lucas, N.M. Markovic, *Science*, 315 (2007) 493.

[6] F. Tao, M.E. Grass, Y. Zhang, D.R. Butcher, J.R. Renzas, Z. Liu, J.Y. Chung, B.S. Mun, M. Salmeron, G.A. Somorjai, *Science*, 322 (2008) 932.

[7] B. Lim, M. Jiang, P.H.C. Camargo, E.C. Cho, J. Tao, X. Lu, Y. Zhu, Y. Xia, *Science*, 324 (2009) 1302.

[8] T. Omori, K. Ando, M. Okano, X. Xu, Y. Tanaka, I. Ohnuma, R. Kainuma, K. Ishida, *Science*, 333 (2011) 68.

[9] E. González, J. Arbiol, V.F. Puntes, *Science*, 334, (2011), 1377.

[10] M. Chen, D. Kumar, C.W. Yi, D.W. Goodman, *Science*, 310, (2005), 291.

[11] L. Kesavan, R. Tiruvalam, M.H. Ab Rahim, M.I. bin Saiman, D.I. Enache, R.L. Jenkins, N. Dimitratos, J.A. Lopez-Sanchez, S.H. Taylor, D.W. Knight, C.J. Kiely, G.J. Hutchings, *Science*, 331 (2011) 195.

[12] G. Kyriakou, M.B. Boucher, A.D. Jewell, E.A. Lewis, T.J. Lawton, A.E. Baber, H.L. Tierney, M. Flytzani-Stephanopoulos, E.C.H. Sykes, *Science*, 335 (2012) 1209.

[13] M.T. Bore, T.L. Ward, R.F. Marzke, A.K. Datye, *J. Mater. Chem,* 15 (2005) 5022 – 5028.

[14] R. Ravishankar, M.M. Li, A. Borgna, *Catalysis Today* 106 (2005) 149 – 153.

[15] K.S.W. Sing, D.H. Everett, R.A.W. Haul, L. Moscou, R.A. Pierotti, J. Rouquerol and T. Siemieniewska, *Pure Appl. Chem.*, 1985, 57, 603.

[16] J.S. Beck, J.C. Vartuli, W.J. Roth, M.E. Leonowicz, K.D. Schmidt, C.T.W. Chu, D.H. Olson, E.W. Sheppard, S.B. McCullen, J.B. Higgins and J.L. Schlenker, *J. Am. Chem. Soc.*, 1992, **114**, 10834.

[17] H.X. Dai, A.T. Bell, E. Iglesia, *Journal of Catalysis* 221 (2004) 491 – 499.

[18] E.V. Kondratenko, M. Cherian, M. Baerns, D. Su, R. Schlogl, X. Wang, I.E. Wachs, *Journal of Catalysis* 234 (2005) 141

[19] A. Ellison, T.L. Overton, L. Benzce, *Journal of the Chemical Society,* Faraday Transactions, 89 (1993) 843

[20] J.Y. Piquemal, E. Briot, M. Vennat, J.M. Brégeault, G. Chottard and J.M. Manoli, *Chem. Commun.*, 1999, 1195.

[21] J.Y. Piquemal, J.M. Manoli, P. Beaunier, A. Ensuque, P. Tougne, A.P. Legrand and J.M. Brégeault, *Micropor. Mesopor. Mater.*, 29, 1999, 291.

[22] W. Zhang, J. Wang, P.T. Tanev and T.J. Pinnavaia, *Chem. Commun.*, 1996, 979.

[33] S.M.K. Airaksinen, A.O.I. Krause, J. Sainio, J. Lahtinen, K.J. Chao, M.O. Guerrero-Perez, M.A. Banares, *Physical Chemistry Chemical Physics* 5 (2003) 4371–4377.

[24] M. Weckhuysen, I.E. Wachs, *Journal of Physical Chemistry* 100 (1996) 437–442.

[25] D.L. Hoang, A. Dittmar, M. Schneider, A. Trunschke, H. Lieske, K.W. Brzezinka, K. Witke, *Thermochimica Acta* 400 (2003) 153–163.

[26] T. Blasco, J.M. Lopez Nieto, A. Dejoz, M.I. Vazque, *J. Catal.* 157, 271–282 (1995).

[27] L.M. Madeira, R.M.M. Aranda, F.J.M. Hodar, J.L.G. Fierro, M.F.Portela, *J. Catal.* 169, 469–479 (1997).

[28] G. Karamullaoglu, T. Dogu, Ind and Engr. Chem. Res. 46 (2007) 7079 – 7086.

[29] R. Serge, *Thermal and Catalytic Processes in Petroleum Refining*, Marcel Dekker, New York, 2003.

[30] H.X. Dai, A.T. Bell, E. Iglesia, *Journal of Catalysis* 221 (2004) 491 – 499.

[31] A.A. Ijagbuji, I.I. Zakharov, T.C. Philips, M.G. Loriya, M.B. Saltzberg, A.B. Tselishtev, R.J. Taylor, B.V. Shevchenko, K.M. Jackson, D.B. Woods, and J.K. Johnson, *Production of olefins via oxidative de-hydrogenation of C_3–C_4 fraction by O_2 over (Cr–Mo)/SiO$_2$*, ISBN 978-3-656-89037-9, Grin Verlag GmbH, Münich, Germany.

[32] D.E. O'Reilly, F.D. Santiago, R.G. Squires, *J. Phys. Chem.* 73(1969) 3172.

17